Basic Mental Math Step 1

Seong R. Kim

Basic Mental Math Step 1

Copyright © 2018 by Seong Ryeol Kim. All rights reserved.

Basic Mental Math Step 1

What's this about?

And why do you need this?

This is designed to help you do practices for your mental math, calculation by heart. So this helps improve your mental math.

Why mental math, though?

Your mental math will help you stay focused when you do math. The more mental math you do, the less you get distracted. More likely to stay focused if doing math mentally.

And this is one of these: Basic Mental Math Step 1, Basic Mental Math Step 2, and so on.

Basic Mental Math Step 1

Each is a practice set, made of examples of simple calculations, very simple. Even just reading them, your mental math improves.

When reading those examples, you are, kind of, forced to do mental math.

With no pressure, though.

Basic Mental Math Step 1

Reading and repeating those examples, you

get to break a number into other numbers

and put them together some different ways.

Sounds hard, doesn't it?

No worries. It begins with some easy ones,

very easy, and gradually, leads you to the

ones bit more involved, bit by bit.

Actually, you're doing calculations mentally in parts. And your mental math grows.

As you get used to each step of practices, Step 1, Step 2, and so forth, your mental math keeps growing, and gets easier, smoothly faster.

Basic Mental Math Step 1

And this is also, particularly for you if you have a difficulty learning math as math anxiety or phobia. So anyway somehow if math is daunting you, this is right for you.

Thus, if you hate math or struggle with it, but you want to learn it right so that you can do some essential math right, smooth, and fast enough, you are looking at the right book right now.

Basic Mental Math Step 1

And handy and wieldy

This book is small enough to carry and has fonts large enough to read, so can be in your pocket, can always be with you, and can be an easy read whenever you have time for it.

Basic Mental Math Step 1

$1 + 0 =$

Basic Mental Math Step 1

$$1 + 0 = 1$$

Basic Mental Math Step 1

$0 + 1 =$

Basic Mental Math Step 1

$0 + 1 = 1$

Basic Mental Math Step 1

0 + 0 =

1 + 0 =

0 + 1 =

Basic Mental Math Step 1

0 + 0 = 0

1 + 0 = 1

0 + 1 = 1

Basic Mental Math Step 1

$1 + 1 =$

Basic Mental Math Step 1

$$1 + 1 = 2$$

$2 + 0 =$

Basic Mental Math Step 1

$2 + 0 = 2$

Basic Mental Math Step 1

$$0 + 2 =$$

$0 + 2 = 2$

Basic Mental Math Step 1

1 + 1 =

2 + 0 =

0 + 2 =

Basic Mental Math Step 1

$1 + 1 = 2$

$2 + 0 = 2$

$0 + 2 = 2$

$2 + 1 =$

Basic Mental Math Step 1

$$2 + 1 = 3$$

Basic Mental Math Step 1

$$1 + 2 =$$

Basic Mental Math Step 1

$$1 + 2 = 3$$

$3 + 0 =$

Basic Mental Math Step 1

$3 + 0 = 3$

$0 + 3 =$

Basic Mental Math Step 1

$$0 + 3 = 3$$

Basic Mental Math Step 1

2 + 1 =

1 + 2 =

$2 + 1 = 3$

$1 + 2 = 3$

Basic Mental Math Step 1

3 + 0 =

0 + 3 =

$3 + 0 = 3$

$0 + 3 = 3$

Basic Mental Math Step 1

$3 + 1 =$

Basic Mental Math Step 1

$$3 + 1 = 4$$

Basic Mental Math Step 1

$1 + 3 =$

Basic Mental Math Step 1

$$1 + 3 = 4$$

Basic Mental Math Step 1

2 + 2 =

Basic Mental Math Step 1

$2 + 2 = 4$

$4 + 0 =$

Basic Mental Math Step 1

$$4 + 0 = 4$$

Basic Mental Math Step 1

0 + 4 =

Basic Mental Math Step 1

$$0 + 4 = 4$$

Basic Mental Math Step 1

$3 + 1 =$

$1 + 3 =$

Basic Mental Math Step 1

$3 + 1 = 4$

$1 + 3 = 4$

Basic Mental Math Step 1

2 + 2 =

4 + 0 =

0 + 4 =

$2 + 2 = 4$

$4 + 0 = 4$

$0 + 4 = 4$

$4 + 1 =$

Basic Mental Math Step 1

$4 + 1 = 5$

Basic Mental Math Step 1

$1 + 4 =$

Basic Mental Math Step 1

$1 + 4 = 5$

Basic Mental Math Step 1

2 + 3 =

Basic Mental Math Step 1

$$2 + 3 = 5$$

Basic Mental Math Step 1

$$3 + 2 =$$

Basic Mental Math Step 1

$3 + 2 = 5$

$5 + 0 =$

Basic Mental Math Step 1

$5 + 0 = 5$

$0 + 5 =$

Basic Mental Math Step 1

$0 + 5 = 5$

$4 + 1 = 1 + 4$

$2 + 3 = 3 + 2$

$5 + 0 = 0 + 5$

Basic Mental Math Step 1

$1 + 1 + 1 + 1 + 1$
$= 1 + 2 + 1 + 1$
$= 1 + 3 + 1$
$= 1 + 4 = 5$

$5 + 1 =$

Basic Mental Math Step 1

$5 + 1 = 6$

$1 + 5 =$

Basic Mental Math Step 1

$$1 + 5 = 6$$

$2 + 4 =$

Basic Mental Math Step 1

$2 + 4 = 6$

$4 + 2 =$

Basic Mental Math Step 1

$4 + 2 = 6$

Basic Mental Math Step 1

3 + 3 =

Basic Mental Math Step 1

$3 + 3 = 6$

6 + 0 =

Basic Mental Math Step 1

$6 + 0 = 6$

$0 + 6 =$

$0 + 6 = 6$

$5 + 1 = 1 + 5$

$4 + 2 = 2 + 4$

$6 + 0 = 0 + 6$

Basic Mental Math Step 1

$$6 = 3 + 3$$
$$= 3 + 2 + 1$$
$$= 2 + 2 + 2$$

$6 + 1 =$

Basic Mental Math Step 1

$$6 + 1 = 7$$

1 + 6 =

Basic Mental Math Step 1

$1 + 6 = 7$

2 + 5 =

Basic Mental Math Step 1

$$2 + 5 = 7$$

Basic Mental Math Step 1

$$5 + 2 =$$

Basic Mental Math Step 1

$$5 + 2 = 7$$

Basic Mental Math Step 1

$4 + 3 =$

$4 + 3 = 7$

$3 + 4 =$

Basic Mental Math Step 1

$$3 + 4 = 7$$

7 + 0 =

Basic Mental Math Step 1

7 + 0 = 7

Basic Mental Math Step 1

$0 + 7 =$

Basic Mental Math Step 1

0 + 7 = 7

Basic Mental Math Step 1

$6 + 1 = 1 + 6$

$2 + 5 = 5 + 2$

$3 + 4 = 4 + 3$

$$7 = 3 + 4$$
$$= 3 + 3 + 1$$
$$= 1 + 2 + 3 + 1$$

Basic Mental Math Step 1

7 + 1 =

$7 + 1 = 8$

1 + 7 =

Basic Mental Math Step 1

$1 + 7 = 8$

$2 + 6 =$

Basic Mental Math Step 1

$$2 + 6 = 8$$

$6 + 2 =$

Basic Mental Math Step 1

$6 + 2 = 8$

Basic Mental Math Step 1

$3 + 5 =$

Basic Mental Math Step 1

$3 + 5 = 8$

Basic Mental Math Step 1

5 + 3 =

Basic Mental Math Step 1

$5 + 3 = 8$

$4 + 4 =$

$4 + 4 = 8$

8 + 0 =

Basic Mental Math Step 1

$8 + 0 = 8$

$0 + 8 =$

Basic Mental Math Step 1

$0 + 8 = 8$

$7 + 1 = 1 + 7$

$6 + 2 = 2 + 6$

$5 + 3 = 3 + 5$

$$8 = 4 + 4$$
$$= 1 + 3 + 3 + 1$$
$$= 2 + 2 + 2 + 2$$

8 + 1 =

$8 + 1 = 9$

$1 + 8 =$

Basic Mental Math Step 1

$1 + 8 = 9$

$2 + 7 =$

Basic Mental Math Step 1

$2 + 7 = 9$

Basic Mental Math Step 1

$7 + 2 =$

$7 + 2 = 9$

$3 + 6 =$

Basic Mental Math Step 1

$$3 + 6 = 9$$

$6 + 3 =$

Basic Mental Math Step 1

$6 + 3 = 9$

Basic Mental Math Step 1

$4 + 5 =$

Basic Mental Math Step 1

$4 + 5 = 9$

$5 + 4 =$

Basic Mental Math Step 1

$$5 + 4 = 9$$

9 + 0 =

Basic Mental Math Step 1

9 + 0 = 9

Basic Mental Math Step 1

0 + 9 =

$0 + 9 = 9$

8 + 1 = 7 + 2

6 + 3 = 5 + 4

7 + 2 = 3 + 6

Basic Mental Math Step 1

$5 + 4 = 8 + 1$

$7 + 2 = 6 + 3$

$= 3 + 3 + 3$

3 + 4 =

2 + 3 =

3 + 6 =

3 + 5 =

Basic Mental Math Step 1

$3 + 4 = 7$

$2 + 3 = 5$

$3 + 6 = 9$

$3 + 5 = 8$

3 + 1 =

7 + 3 =

3 + 8 =

9 + 3 =

Basic Mental Math Step 1

$3 + 1 = 4$

$7 + 3 = 10$

$3 + 8 = 11$

$9 + 3 = 12$

4 + 5 =

2 + 4 =

3 + 4 =

4 + 6 =

Basic Mental Math Step 1

$4 + 5 = 9$

$2 + 4 = 6$

$3 + 4 = 7$

$4 + 6 = 10$

4 + 7 =

8 + 4 =

9 + 4 =

1 + 4 =

Basic Mental Math Step 1

4 + 7 = 11
8 + 4 = 12
9 + 4 = 13
1 + 4 = 5

$5 + 7 =$

$8 + 5 =$

$9 + 5 =$

$5 + 4 =$

$5 + 7 = 12$

$8 + 5 = 13$

$9 + 5 = 14$

$5 + 4 = 9$

3 + 5 =

5 + 2 =

6 + 5 =

1 + 5 =

$3 + 5 = 8$

$5 + 2 = 7$

$6 + 5 = 11$

$1 + 5 = 6$

3 + 7 =

7 + 2 =

6 + 7 =

7 + 5 =

Basic Mental Math Step 1

$3 + 7 = 10$

$7 + 2 = 9$

$6 + 7 = 13$

$7 + 5 = 12$

1 + 7 =

7 + 4 =

8 + 7 =

7 + 9 =

Basic Mental Math Step 1

$1 + 7 = 8$

$7 + 4 = 11$

$8 + 7 = 15$

$7 + 9 = 16$

Basic Mental Math Step 1

8 + 7 =

4 + 8 =

8 + 1 =

8 + 9 =

$8 + 7 = 15$

$4 + 8 = 12$

$8 + 1 = 9$

$8 + 9 = 17$

Basic Mental Math Step 1

8 + 2 =
3 + 8 =
8 + 5 =
8 + 6 =

$8 + 2 = 10$

$3 + 8 = 11$

$8 + 5 = 13$

$8 + 6 = 14$

6 + 7 =

4 + 6 =

6 + 3 =

2 + 6 =

6 + 7 = 13
4 + 6 = 10
6 + 3 = 9
2 + 6 = 8

$6 + 8 =$

$5 + 6 =$

$1 + 6 =$

$6 + 9 =$

$6 + 8 = 14$

$5 + 6 = 11$

$1 + 6 = 7$

$6 + 9 = 15$

Basic Mental Math Step 1

$6 + 2 =$

$5 + 2 =$

$2 + 3 =$

$2 + 9 =$

$6 + 2 = 8$

$5 + 2 = 7$

$2 + 3 = 5$

$2 + 9 = 11$

8 + 2 =

4 + 2 =

2 + 7 =

2 + 1 =

Basic Mental Math Step 1

$8 + 2 = 10$

$4 + 2 = 6$

$2 + 7 = 9$

$2 + 1 = 3$

Basic Mental Math Step 1

1 + 2 =

4 + 1 =

5 + 1 =

1 + 7 =

Basic Mental Math Step 1

$1 + 2 = 3$

$4 + 1 = 5$

$5 + 1 = 6$

$1 + 7 = 8$

1 + 8 =

6 + 1 =

9 + 1 =

1 + 3 =

Basic Mental Math Step 1

1 + 8 = 9

6 + 1 = 7

9 + 1 = 10

1 + 3 = 4

9 + 8 =

6 + 9 =

9 + 4 =

9 + 3 =

$9 + 8 = 17$

$6 + 9 = 15$

$9 + 4 = 13$

$9 + 3 = 12$

9 + 2 =

5 + 9 =

9 + 7 =

9 + 1 =

Basic Mental Math Step 1

$9 + 2 = 11$

$5 + 9 = 14$

$9 + 7 = 16$

$9 + 1 = 10$

www.ingramcontent.com/pod-product-compliance
Lightning Source LLC
Chambersburg PA
CBHW020657220526
45464CB00001B/479